Ian Glenton

Artificial Intelligence: Between Progress and its Shadows

Ian Glenton

Artificial Intelligence: Between Progress and its Shadows

Learning to be more cautious of new technology

ScienciaScripts

Imprint

Cover image: www.ingimage.com

This book is a translation from the original published under ISBN 978-620-0-01424-5.

Publisher:
Sciencia Scripts
is a trademark of
Dodo Books Indian Ocean Ltd. and OmniScriptum S.R.L publishing group

120 High Road, East Finchley, London, N2 9ED, United Kingdom
Str. Armeneasca 28/1, office 1, Chisinau MD-2012, Republic of Moldova, Europe
Managing Directors: Ieva Konstantinova, Victoria Ursu
info@omniscriptum.com

Printed at: see last page
ISBN: 978-620-8-53661-9

Artificial Intelligence: Between Progress and its Shadows

Author:

Ian Etienne Glenton Montenegro

Table of Contents

Introduction

Artificial intelligence (AI) has burst into all aspects of our daily lives, from the simplest to the most complex applications, promising a future full of technological advances that seem to have no limits. In this context, AI is presented as the engine of unstoppable progress, a tool capable of transforming society, the economy and the way we relate to the world. But behind this silicon dream lies a more complex and worrying reality.

This five-part tour invites us to reflect on the dark side of AI, a world where invisible pollution, digital waste, and the exploitation of human and natural resources are beginning to show their traces. As AI advances with the promise of a more efficient future, it is essential to question its consequences, both on the environment around us and on the social structures we have built.

In this journey, we will explore how data clouds and the energy consumed by data centers are a heavy environmental burden, and how technological advancement has brought with it not only progress, but also new risks, such as ethical biases seeping into algorithms. However, the reflection does not stop at pessimism, but also seeks answers towards a sustainable future, where innovation can go hand in hand with social and environmental responsibility.

Ultimately, this tour aims to make us question: Are we willing to pay the price of progress, often at the cost of the planet's health and social equity?

Is it possible to strike a balance between artificial intelligence and a future that is ethical, sustainable and truly human?

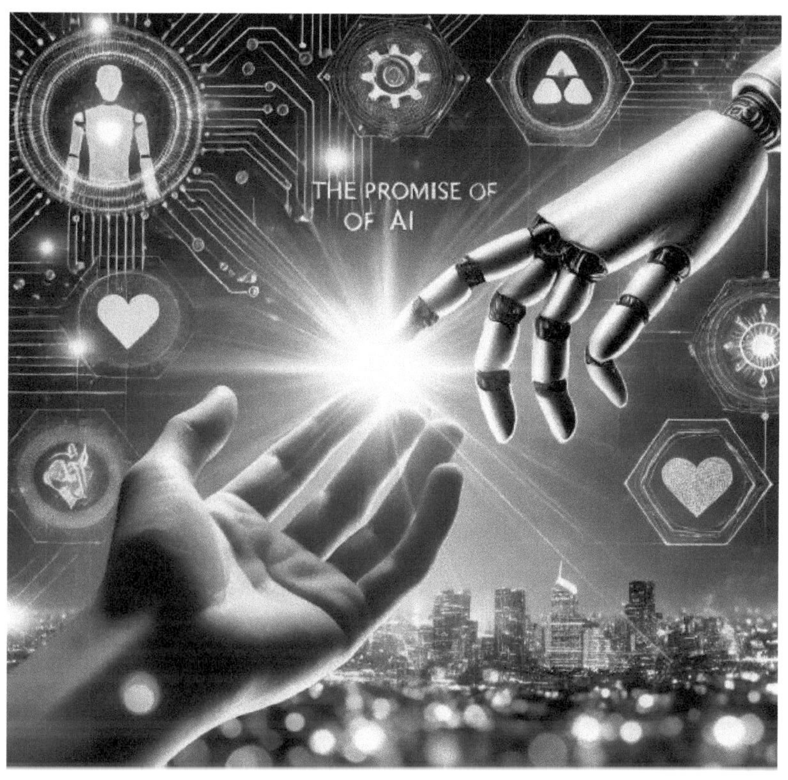

Chapter 1: The Promise of AI

Silicon Dreams: AI as the Engine of the Future

The rise of artificial intelligence (AI) is shaping a new era in human history, characterized by technological advances that are redefining the boundaries of what is possible. From systems capable of performing complex tasks to generative models such as chatbots, AI is emerging as a powerful tool for transforming society. However, its development raises essential questions about the ethical, economic and social implications of this technology.

The engine of progress

AI is often described as **"humanity's latest invention,"** as its potential to solve complex problems could radically change the world we live in (Bostrom, quoted in Rísbel Magazine, 2023). This optimism is complemented by predictions such as those of Ray Kurzweil, who foresees that by 2045 machine intelligence will not only equal, but surpass human intelligence, leading to what is known as the Technological Singularity. This future poses both hope and uncertainty.

In areas such as medicine, advanced algorithms are already helping to diagnose diseases more accurately than human doctors. On the other hand, entire industries are embracing automation to improve efficiency and reduce costs. However, the rapid advancement of AI also raises concerns. Elon Musk, for example, warns that this development could bring significant risks if not properly regulated, comparing it to a technology as powerful as fire: useful, but dangerous if not handled correctly (Rísbel Magazine, 2023).

Benefit and risk: an ambivalent relationship

AI has an intrinsic duality. According to François Chollet, its impact depends on how we use it: as a tool of creation or destruction. This approach highlights the need for ethical and responsible management. Furthermore, Kate Crawford argues that AI is neither truly "intelligent" nor "artificial," as it reflects human biases and aspirations. This underscores the importance of designing systems that do not perpetuate social inequalities or ethical errors (Entrepreneurs News, 2023).

Future and responsibility

To ensure that AI benefits society, we must adopt an ethical perspective. Fei-Fei Li, an expert in artificial intelligence, states that AI **"is not just about algorithms and technology, but also about ethics and human values"** (Rísbel Magazine, 2023). This implies that developers and users must work together to integrate ethical principles at all stages of its creation and use.

Ultimately, the silicon dream that AI promises will only be fulfilled if we meet its challenges with creativity, responsibility and vision. The decisions we make today will determine whether AI becomes a driver of sustainable progress or a destabilizing factor for our society.

Algorithmic Revolutions: Beyond the Human

Artificial intelligence (AI) represents an unprecedented transformation in the ability of machines to emulate, complement and surpass human abilities. From automating complex processes to continuous learning, this technology is redefining not only industries, but also the boundaries of the human.

Transformation of industries

AI has revolutionized key sectors such as healthcare, manufacturing and transportation. In medicine, it enables more accurate diagnoses through the analysis of images and clinical data, while in the automotive industry, autonomous vehicles are redefining mobility and road safety. In addition, their impact on customer service through virtual assistants improves the user experience by providing personalized and immediate responses (Tec de Monterrey, 2024).

In research and development, AI optimizes innovation by processing large volumes of data with speed and accuracy unattainable by humans. This has enabled significant advances in sectors such as pharmaceuticals, where drug discovery is accelerated and clinical trials are optimized (Unite.AI, 2024).

Challenges and opportunities

Despite its benefits, AI poses ethical and social challenges. For example, the displacement of workers in repetitive roles creates tensions in the labor market. However, it also creates new opportunities in areas such as software development, the digital economy and applied creativity, where human-machine collaboration amplifies creative potential (Inter-American Development Bank, 2024).

In addition, the ethical use of AI is critical. AI systems must be trained with unbiased data and used under principles that respect privacy and human dignity. This requires constant monitoring to avoid bias and adverse outcomes (UNAM Gazette, 2024).

Artificial Intelligence: A Seemingly Clean Progress

Artificial Intelligence (AI) is often perceived as an elegant solution to present and future problems, promoted as a tool capable of improving efficiency and sustainability in various industries. However, its true environmental impact defies this narrative, revealing a significant cost in terms of energy, natural resources and polluting emissions.

The energetic weight of AI

Data centers, essential for training and operating AI systems, consume massive amounts of electrical power. According to a report in the journal *Nature*, these facilities account for between 1% and 2% of global electricity consumption, and it is projected that this figure could triple in the coming decades due to increased demand for AI and cloud services. Training a large model, such as those used for advanced natural language processing tasks, can emit up to 284 metric tons of carbon dioxide (Strubell et al., 2019), comparable to the annual environmental impact of 60 average households in industrialized countries.

In addition, energy use does not always come from clean sources. Despite advances in renewable energy, much of the data centers are still powered by fossil fuels. In regions such as the United States and China, which are home to some of the world's largest data centers, coal and natural gas are still predominant sources of energy.

Water consumption and critical minerals

A less discussed, but equally important impact is water consumption. Data centers use water to cool their systems, a process that can be unsustainable in drought-prone regions. For example, data centers in Silicon Valley are estimated to consume millions of gallons of water annually, exacerbating the water crisis in areas already affected by climate change.

In parallel, the manufacture of the hardware needed to support AI technologies depends on critical minerals such as cobalt, lithium and nickel. These materials are mined primarily in countries in the global south under environmentally and socially questionable conditions. Mining often degrades local ecosystems and generates social conflict over control of resources.

Does the apparent sustainability pay off?

While AI has proven useful in sustainable projects, such as optimizing vehicular traffic or designing predictive models for climate change, these applications do not fully balance their negative environmental impact. According to a report by the *International Energy Agency (IEA)*, increasing the energy efficiency of data centers is crucial, but it will not be enough unless strict global policies are adopted to regulate and minimize resource consumption.

In addition, AI-based solutions can amplify environmental problems. For example, the rise of cryptocurrencies and NFTs, driven in part by similar technologies, has resulted in an exponential increase in energy consumption due to intensive digital mining processes.

The future of truly clean progress

For AI to fulfill its promise as a tool for sustainable progress, it is essential to adopt a responsible and innovative approach. This includes:

1. *Adoption of renewable energy:* Leading AI companies, such as Google and Microsoft, have begun to commit to carbon-neutral operations, but these initiatives must become a global standard.
2. *Recycling and reuse of components:* Encourage circular economy in the manufacture of technology hardware to minimize dependence on extractive mining.
3. *Greater transparency and regulation:* Require technology companies to publish full environmental audits on the impact of their AI operations and align with global targets such as net-zero emissions.
4. *Promote resource efficiency:* Design lighter and less energy-intensive models that can operate with lower computational demand without compromising quality.

Chapter 2: Invisible Pollution

Data Clouds, Carbon Footprinting

In today's digital era, cloud services and data centers have revolutionized the way we store, process and share information. However, this technological advancement has a significant environmental impact that often goes unnoticed. The "invisible pollution" generated by digital infrastructures, especially data centers, results in a considerable increase in carbon emissions. This paper explores the environmental impact of cloud services in terms of energy consumption, carbon footprint and sustainability, based on recent research and reliable data.

Data center energy consumption

Data centers are the cornerstones of cloud computing. These facilities, which house servers responsible for processing and storing massive amounts of data, require a significant amount of electricity. According to a report by the International Energy Agency (IEA), data centers accounted for approximately **1% of global electricity consumption in 2022** (IEA, 2022). Although this percentage may seem low, the demand for cloud services is growing exponentially, which could double this consumption in the coming decades if sustainable solutions are not implemented.

In addition to direct energy consumption, a large portion of the energy used by data centers goes to cooling systems. Servers generate heat while operating, and maintaining an adequate temperature is crucial to prevent failures and ensure optimal performance. However, traditional cooling systems are highly inefficient and contribute to excessive electricity consumption.

The carbon footprint of the cloud

The environmental impact of cloud services is not limited to electricity consumption, but also involves the emission of greenhouse gases, known as carbon footprint. Although many technology companies have adopted policies to reduce their environmental impact, the constant growth of digital demand hinders these efforts.

For example, a study by Anders Andrae and Tomas Edler (2015) estimated that information and communication technologies (ICT), including networks, devices and data centers, could account for up to **14% of global carbon emissions by 2040**, if no significant action is taken. This is comparable to emissions from the entire transportation industry, highlighting the severity of the problem.

Strategies towards sustainability

Against this backdrop, technology companies and governments are implementing various strategies to mitigate the environmental impact of cloud services. One of the most promising solutions is the transition to renewable energy sources. Technology giants such as Google, Amazon and Microsoft have invested significantly in solar and wind energy projects to power their data centers. For example, Google announced in 2020 that it had achieved a net carbon balance since its founding in 1998 and committed to operate entirely on carbon-free energy by 2030 (Google, 2020).

Another important measure is the optimization of energy efficiency through technological innovations. The most advanced data centers use artificial intelligence (AI) to manage energy consumption and reduce waste. For example, Google implemented AI systems to optimize cooling in its data centers, achieving a 30% reduction in cooling-related energy costs (DeepMind, 2018).

The role of the end user

While technology companies play a crucial role in reducing the carbon footprint of the cloud, end users also have an important responsibility. Practices such as deleting unnecessary emails, reducing cloud storage usage, and unplugging devices when they are not in use may seem insignificant, but they have a significant cumulative impact. According to a study published by The Shift Project (2019), online video traffic accounts for nearly **60% of global data traffic**, and its carbon footprint is equivalent to that of all of Spain in one year. Reducing the consumption of unnecessary digital content is therefore an effective measure at the individual level.

The Dark Side of Data Centers

In an increasingly digitized world, data centers are at the heart of the global technology infrastructure. These gigantic server complexes not only store and process data, but are also essential for the operation of cloud services, social networks, video streaming, e-commerce and more. However, behind their efficiency and ability to support modern life lies a "dark side" that generates enormous environmental, social and ethical costs. This essay explores the problems associated with data centers, including their skyrocketing energy consumption, environmental impact, privacy risks, and concentration of power in the hands of a handful of technology corporations.

The enormous energy consumption

One of the most worrying aspects of data centers is the massive amount of energy they consume to operate. According to the International Energy Agency (IEA), data centers consumed around **200 terawatt-hours (TWh)** in 2022, which is roughly equivalent to the total electricity consumption of a country like Spain (IEA, 2022). This energy consumption not only powers the servers that process data, but also the cooling systems needed to maintain optimal temperatures in these facilities.

The problem is that much of this energy comes from non-renewable sources, such as coal and natural gas, which contributes to greenhouse gas emissions. In particular, data centers located in regions where fossil energy sources predominate have a much larger carbon footprint. For example, in countries such as China, where coal is still a major source of electricity, data centers generate significantly higher carbon emissions compared to those operating in Europe or North America, where the use of renewable energy is more widespread (Andrae & Edler, 2015).

Environmental impact: beyond energy

The environmental impact of data centers is not limited to energy consumption. The construction and maintenance of these facilities require enormous amounts of natural resources, including water, metals and land.

One of the most critical issues is the intensive use of water for cooling systems. According to a report by the **U.S. Department of Energy**, some data centers can consume up to **millions of liters of water per day** to maintain adequate temperatures, putting significant pressure on local water supplies, especially in drought-stricken regions (DOE, 2022).

In addition, the manufacture of servers and technological equipment requires the extraction of rare metals such as lithium, cobalt and nickel. Mining of these materials not only has devastating environmental impacts, such as deforestation and river pollution, but is also associated with human rights violations, including child labor and labor exploitation in developing countries (Amnesty International, 2016).

Risks to data privacy and security

The dark side of data centers is not limited to the environmental sphere. The massive storage of data in these facilities poses considerable risks to users' privacy and security. Data leaks, cyberattacks and misuse of personal information are constant concerns in a world where technology companies handle unimaginable amounts of sensitive data.

For example, in 2021, a cyberattack on a Facebook data center exposed the personal data of more than 530 million users, including phone numbers and email addresses (BBC News, 2021). This incident underscores the inherent vulnerability of these infrastructures, which become prime targets for hackers due to the wealth of information they contain.

In addition, data storage in the cloud, while convenient, raises ethical questions about the control of information. Companies such as Amazon, Microsoft, and Google concentrate a large share of the global market for cloud services, giving them disproportionate power over the data of individuals, businesses, and governments. This centralization raises concerns about technological monopolies and the potential manipulation of data for political or commercial purposes (Zuboff, 2019).

Concentration of power and global inequalities

The dominance of a few technology giants in the data center and cloud computing sector also has economic and social implications. Companies such as Amazon Web Services (AWS), Microsoft Azure and Google Cloud control more than **60% of the global market for cloud services** (Statista, 2023), allowing them to set the rules of the game and limit the competitiveness of smaller companies.

This digital monopoly perpetuates global inequalities. On the one hand, developing countries are increasingly dependent on the technological infrastructures of foreign companies, reducing their digital sovereignty. On the other hand, the economic and technological benefits of data centers are often concentrated in developed countries, while environmental and social costs, such as pollution and labor exploitation, are outsourced to developing countries.

Towards a more sustainable and ethical future

To address the "dark side" of data centers, it is essential to implement sustainable and ethical strategies that balance the benefits of technology with the need to protect the environment and human rights. Some of the solutions include:

1. **Transition to renewable energy:** Companies such as Google and Microsoft are already investing in wind and solar energy

projects to power their data centers. According to Google, its goal is to operate entirely on carbon-free energy by 2030 (Google, 2020).

2. **Optimizing energy efficiency:** Using artificial intelligence to optimize energy consumption and cooling systems can significantly reduce environmental impact. For example, Google has managed to decrease **energy consumption in its cooling systems** by **40%** through the use of AI (DeepMind, 2018).

3. **Stricter regulations:** Governments should implement policies that force technology companies to comply with environmental and labor standards, in addition to encouraging transparency in data management and fair competition in the marketplace.

4. **Empowering users:** Finally, consumers can contribute to change by adopting responsible digital practices, such as reducing unnecessary cloud storage and demanding greater sustainability and data protection from companies.

Digital Garbage: Where Do Old Algorithms End?

Currently, much of the conversation about sustainability and digital pollution revolves around the energy consumption of data centers, the carbon footprint of information technology and the accumulation of discarded electronic devices. However, there is a less visible, but equally troubling aspect: "digital waste". This concept refers not only to obsolete or useless data, but also to algorithms, systems and applications that, once discarded, remain stored on servers or devices without a clear purpose, contributing to the unnecessary consumption of resources. This essay explores the impact of this form of digital pollution, its ethical and environmental implications, and possible solutions to better manage the life cycle of algorithms.

The concept of digital waste

The term "digital waste" is not limited to user-generated content such as e-mails, duplicate files or unused applications. It also includes algorithms, software and systems that, once replaced by more advanced versions, remain stored in servers and data centers. Although these algorithms are no longer in active use, their storage continues to consume energy and digital space.

According to a study by The Shift Project (2019), inactive data storage accounts for a significant portion of global digital traffic and energy consumption. This is because the servers that store this "digital garbage" must be kept operational, which involves continuous use of electricity and cooling systems. This phenomenon is comparable to keeping unnecessary physical files in large warehouses, but with a much greater environmental impact.

The life cycle of algorithms

Algorithms, like any technology, have a life cycle. Initially, they are developed to solve specific problems or improve processes. However, over time, they become obsolete due to the emergence of new technologies, changes in user needs or advances in artificial intelligence. Once an algorithm is no longer useful, it is not always eliminated; in many cases, it is archived or simply abandoned on servers.

This abandonment raises several ethical questions. For example, who is responsible for managing obsolete algorithms? Should they be disposed of entirely or preserved as part of a "digital archive" of humanity? While some argue that old algorithms can be reused or studied to improve future development, others point out that keeping them in storage indefinitely is an unsustainable practice.

Environmental impact of digital waste

The storage of dormant algorithms and data contributes to digital pollution in several ways:

1. *Unnecessary energy consumption:* As mentioned by the International Energy Agency (IEA, 2022), data centers already account for a significant percentage of global electricity consumption. Storing obsolete algorithms and other unnecessary data increases this energy burden without providing real value.
2. *Use of material resources:* Although digital waste is intangible, its storage requires physical infrastructure, such as servers and hard disks. The manufacture of this equipment depends on the extraction of natural resources, such as rare metals, the exploitation of which has a high environmental cost.
3. *E-waste generation:* servers and devices that store dormant algorithms and data eventually wear out and are discarded, contributing to the e-waste problem. According to a report by the

United Nations (UN, 2020), the world generated **53.6 million tons of e-waste in 2019**, and only 17.4% was properly recycled.

Ethical implications of deprecated algorithms

In addition to the environmental impact, digital waste raises ethical issues related to privacy, security and transparency. Old algorithms, even if no longer in use, may contain sensitive or historical data that, if not properly managed, could be vulnerable to unauthorized access.

For example, in 2019, a marketing company's server exposed a database of old algorithms and logs containing personal information of millions of users (TechCrunch, 2019). This incident demonstrates that the lack of digital waste management not only creates environmental problems, but also puts privacy and data security at risk.

In addition, the existence of obsolete algorithms raises questions about corporate responsibility. Many companies stockpile these systems without considering their long-term impact, reflecting a lack of planning in technology lifecycle management.

Possible solutions for digital waste

To address the problem of digital waste, it is necessary to implement strategies that promote sustainability and ethics in the management of data and algorithms. Some solutions include:

1. *Active algorithm lifecycle management:* Companies and developers should adopt policies to responsibly dispose of obsolete algorithms or archive them in more efficient systems. This includes

implementing processes to regularly evaluate the usefulness of archived systems.

2. *Storage optimization:* Migrating inactive data and algorithms to low-energy storage infrastructures could significantly reduce the environmental impact. For example, technologies such as cold storage, which uses less energy for infrequently accessed data, could be a viable solution (Google, 2020).

3. *Transparency and regulation:* Governments and organizations should establish clear regulations on the management of obsolete algorithms and data. This includes requirements for secure data disposal and minimizing unnecessary storage.

4. *Education and awareness:* End users also have a role to play. Reducing the production of digital waste, such as deleting unnecessary files and applications, can help reduce the overall load on digital systems.

Exploitation and Imbalance: Human and Natural Resources

In the technological era, progress and innovation have become key drivers of the global economy. However, behind this apparent prosperity lies a less visible reality: the exploitation of human and natural resources that underpins the technological and digital infrastructure. From the extraction of rare metals for electronic devices to the precarious working conditions in factories and data centers, the imbalance between the profits made by large corporations and the costs borne by workers and the environment poses serious ethical, social and environmental challenges. This essay analyzes the relationship between the exploitation of human and natural resources in the technology industry, the impact it generates and possible solutions to mitigate this imbalance.

Exploitation of natural resources: the hidden cost of technology

Modern technology depends on materials such as lithium, cobalt, nickel and tantalum, known as "rare metals". These resources are essential for the manufacture of batteries, semiconductors and other electronic components. However, the extraction of these materials comes at a high environmental and social cost. According to a report by the United Nations (UN, 2020), rare metal mining contributes to deforestation, water pollution and soil degradation, severely affecting local ecosystems.

An emblematic case is cobalt mining in the Democratic Republic of Congo (DRC), which produces more than **70% of the world's cobalt supply**. Although cobalt is critical for batteries in electronic devices and electric vehicles, its extraction has been linked to serious human rights violations, including child labor and unsafe working conditions (Amnesty International, 2016). Miners, often poorly paid, work in dangerous conditions and without adequate protective equipment, while multinational companies make huge profits.

In addition, demand for these materials continues to grow due to increased consumption of electronic devices and the rise of renewable energy, intensifying pressure on natural resources. According to a World Bank report (2020), demand for minerals such as lithium and cobalt could increase by more than **500% by 2050** if sustainable alternatives are not developed.

Exploitation of human resources: technology supply chains

The imbalance is also evident in the working conditions of workers who manufacture electronic devices and maintain digital infrastructure. From assembly factories in Asia to data centers in the West, employees often face long hours, low wages and dangerous work environments.

A prominent example is the case of Foxconn, one of the largest electronics manufacturing companies in the world, which produces devices for brands such as Apple, Microsoft and Amazon. In 2010, Foxconn faced global scrutiny after several workers committed suicide due to extreme working conditions, which included grueling shifts, constant surveillance, and lack of emotional support (Chan et al., 2013). Although technology companies have taken steps to improve working conditions in response to these incidents, exploitative practices persist in many parts of the supply chain.

On the other hand, workers in data centers and digital platforms also face challenges. In data centers, which are responsible for storing and processing internet data, employees perform physically strenuous tasks, such as moving heavy servers and maintaining cooling systems, often in extremely hot conditions and with little protection. Likewise, content moderators on platforms such as Facebook and YouTube face severe psychological stress due to constant exposure to graphic and disturbing content, while receiving low wages and little attention to their mental well-being (Newton, 2019).

The imbalance in the distribution of benefits

Although human and natural resources are fundamental to the functioning of the technology industry, the benefits are unevenly distributed. Large technology companies generate billions of dollars in revenue, while workers and communities affected by resource extraction receive a tiny fraction of these profits.

For example, in 2022, Apple reported revenues of more than **$394 billion**, while many of the workers in the factories that produce its devices earn less than **$2 per hour** (Apple, 2022). This imbalance reflects a global economic structure that prioritizes corporate profit over social and environmental justice.

In addition, local communities in areas of rare metal extraction often do not benefit from the economic development associated with mining. Instead, they face degradation of their land, pollution of their water sources and loss of traditional livelihoods. This imbalance perpetuates poverty and inequality in many regions of the world.

Environmental and social impact: the cost of imbalance

The imbalance in the exploitation of human and natural resources has devastating consequences for both the environment and local communities. Overexploitation of natural resources not only depletes ecosystems, but also contributes to climate change. According to the Intergovernmental Panel on Climate Change (IPCC, 2022), mining and technology manufacturing activities are responsible for a significant portion of global carbon emissions.

On a social level, labor exploitation in the technology industry perpetuates cycles of poverty, inequality and abuse, especially affecting the most vulnerable communities. In addition, the lack of adequate regulation in supply chains allows companies to avoid legal and ethical responsibilities, perpetuating unsustainable practices.

Towards a fairer and more sustainable technology industry

To address the imbalance in the exploitation of human and natural resources, it is necessary to implement solutions that promote social justice , environmental sustainability and corporate responsibility. Some of these solutions include:

1. *Transparency in supply chains*: Technology companies need to be more transparent about the origin of the materials they use and the labor conditions in their supply chains. Certifications such as the Responsible Minerals Initiative (RMI) can help ensure more ethical practices.
2. *Investing in recycling and circular economy:* Recycling rare metals and reusing electronic devices can reduce reliance on mining. According to a UN report (2020), only **17.4% of e-waste** is recycled, representing a significant opportunity to improve sustainability.
3. *Improving working conditions:* Companies must ensure fair wages, safe working conditions and psychological support for workers throughout the supply chain. This includes factory employees as well as content moderators and other roles in the digital economy.

4. *Government regulations:* Governments should implement policies that force corporations to comply with environmental and labor standards, as well as impose penalties for abusive practices.

5. *Consumer empowerment:* Consumers have the power to demand more sustainable and ethical practices from companies, choosing brands that prioritize social and environmental justice.

Chapter 3: Consequences and Risks

The Age of Planned Obsolescence: Hardware and Waste

Planned obsolescence is an economic and technical phenomenon that has established itself over the years as a central component in the manufacture and design of products, especially in consumer technology. This business strategy refers to the practice of designing products with a limited useful life, in order to influence the consumer purchase cycle and encourage the continuous consumption of new products. This essay will explore the consequences and risks associated with planned obsolescence, with special emphasis on technological hardware and its effects on e-waste generation.

Definition and Origins of Planned Obsolescence

Programmed obsolescence, a term popularized in the 1930s, refers to the practice of deliberately reducing the useful life of a product to increase sales of new versions. This practice has been observed especially in the technology sector, where products, such as cell phones, computers and household appliances, are often designed to become inoperable or less efficient after a predictable period of time (Slade, 2006). The concept of planned obsolescence is rooted in the need to maintain a constant demand for new products and ensure profit maximization through the rapid renewal cycle.

Environmental Consequences: Increase in Electronic Waste

One of the most negative effects of planned obsolescence is the acceleration of electronic waste or e-waste generation. Electronic devices are often discarded when they no longer function properly, often due to internal component failures that cannot be easily repaired, such as non-replaceable batteries or obsolete chips (Babbitt, 2017). This e-waste disposal generates significant environmental impact, as many of these products contain hazardous materials such as lead, mercury, and cadmium, which, when released into the environment, can contaminate water, air, and soil (Leung et al., 2015).

Despite the fact that most e-waste is recyclable, recycling systems are still insufficient, meaning that a large portion of discarded devices end up in landfills or informal recycling markets, where working conditions are precarious and environmental safety is compromised (Baldé et al., 2017).

Social and Economic Impact

In addition to the environmental repercussions, planned obsolescence also has a considerable economic and social impact. For consumers, the constant need to replace products can generate a considerable financial burden, especially in countries with more vulnerable economies. Planned obsolescence also contributes to the technological gap between developed and developing countries, as cheaper products tend to have shorter life cycles, making access to durable and sustainable technologies more difficult (Akenji, 2014).

From a business perspective, planned obsolescence can be seen as a strategy to maintain competitiveness in the market, but in the long term it can lead to an overproduction of products that, when not properly recycled, generate more problems than solutions. This also creates a paradox, where consumers are not only forced to buy products more frequently, but also face the side effects of a market saturated with discarded products (Cohen, 2017).

Regulation and Global Response

Faced with these problems, various governments and international organizations have begun to take measures to combat programmed obsolescence and mitigate its effects. The European Union, for example, has implemented legislation such as the Waste Electrical and Electronic Equipment (WEEE) Directive, which establishes strict rules on the recycling and disposal of e-waste, promoting the reuse of components and the design of more sustainable products (European Commission, 2018).

However, policies are still not uniform globally, and the implementation of effective standards remains a challenge. In many countries, regulations are limited or not adequately enforced, perpetuating the cycle of mass production of obsolete products and the growing accumulation of e-waste (Baldé et al., 2017).

Innovation in Design and the Circular Model

To address the risks of planned obsolescence, some players in the technology industry have begun to adopt circular economy principles. This model seeks to extend the useful life of products, reduce waste generation and maximize the reuse of materials (Stahel, 2016). Companies such as Fairphone have designed modular mobile devices, which allow users to replace individual parts, such as batteries and screens, thus extending the useful life of the product. This approach not only contributes to reducing e-waste, but also promotes a more responsible and sustainable consumption culture.

Biases and Errors: Ethical Pollution in AI

Artificial intelligence (AI) has emerged as one of the most transformative technologies of the modern era, with applications ranging from healthcare to social justice. However, as AI becomes increasingly integrated into various aspects of everyday life, significant concerns also arise about the biases and errors inherent in AI systems. This phenomenon, which has been referred to as "ethical pollution," poses risks to both individuals and society as a whole. This essay explores the consequences and risks of biases and errors in AI, as well as the ethical and social implications of these phenomena.

Definition of Artificial Intelligence Biases

Bias in AI occurs when algorithms make decisions or predictions that reflect unfair biases, often based on biased data or model programming. These biases may be present in the training data feeding the AI, which may be biased due to historical, social, or cultural factors. In other cases, biases may arise from errors in algorithm design or data interpretation. According to a Harvard University report (2020), AI systems can amplify existing inequalities if biases in their data or algorithms are not properly managed, which can have devastating effects on minority groups.

Causes of AI Biases

Biases in AI systems are generally the result of several factors, including biased data and human decisions during the development process. Training data is critical for an AI model to learn patterns and make accurate predictions. If this data is incomplete, outdated, or representative of only one group, the model will learn to replicate those deficiencies. For example, in the area of hiring, if an AI algorithm is trained on historical data that favors a specific demographic group (e.g., white males), the system will tend to perpetuate that trend, discriminating against other groups, such as women or ethnic minorities (Dastin, 2018).

In addition, decisions made during model design and programming may also introduce biases. The selection of features or variables used to train AI may reflect biases of the designers, even if the designers are not aware of them. For example, if an AI algorithm is used to predict the risk of recidivism in the criminal justice system, it may be based on historical data that reflect racial disparities in convictions, which may generate unfair decisions toward certain racial or ethnic groups (Angwin et al., 2016).

Errors in AI and "Ethical Pollution".

The concept of "ethical pollution" refers to the adverse effects that arise when AI systems, being permeated by biases and errors, violate fundamental ethical principles, such as fairness, justice and transparency. These errors have not only practical, but also philosophical consequences, as they can erode public confidence in the technology. Ethical pollution" occurs when systematic errors in AI have a negative impact on individuals and communities, primarily affecting those groups that have historically been marginalized or underserved.

An emblematic case of ethical pollution in AI was the use of algorithms in the criminal justice system in the United States, such as the COMPAS (Correctional Offender Management Profiling for Alternative Sanctions) risk assessment software, which was used to predict the likelihood of recidivism of offenders. A study by ProPublica (2016) revealed that this system had racial biases, as it was more likely to erroneously predict African Americans committing violent crimes, while whites were less likely to receive these predictions. These types of errors not only affect criminal justice, but perpetuate systemic discrimination in society.

Social and Economic Impact of Biases

Biases in AI have profound implications at both the social and economic levels. At the societal level, these biases reinforce existing inequalities and perpetuate stereotypes, which can exacerbate divisions in society. For example, in healthcare, AI algorithms used to forecast healthcare needs can, if not handled properly, discriminate against the most vulnerable populations. A study by Obermeyer et al. (2019) showed that an algorithm used in the U.S. healthcare system , which predicted which patients needed more care, favored wealthier patients, even though poorer and racially diverse patients suffered greater morbidity.

Economically, biases in AI can distort labor markets, hiring processes and job opportunities, especially affecting disadvantaged groups. In addition, errors in algorithms can have a significant cost in terms of efficiency, as they can make incorrect decisions based on erroneous or biased data. This leads to misuse of resources and inefficient decision making.

Ethical Solutions and Approaches to Mitigate Biases

As AI develops, several approaches have emerged to mitigate the risks of bias and error. One of the most important approaches is "algorithmic auditing," which involves regularly evaluating algorithms and data to identify and correct biases. This audit can be performed by external auditors or by multidisciplinary teams within organizations, and aims to ensure that algorithms do not unfairly favor certain groups or individuals (Eubanks, 2018).

Another key approach is diversity in AI development. Incorporating diverse teams of developers, designers, and ethicists can help identify implicit biases that might otherwise be overlooked. In addition, ethical frameworks and principles are being developed for AI, such as the principles of fairness, accountability, and transparency, that can guide the design and implementation of these systems in ways that minimize the risks of ethical pollution.

Collisions Between Nature and Technology

The rapid advance of technology has transformed every aspect of modern society, from the way we communicate to how we manage our natural resources. However, the increasing integration of technology into everyday life has also generated a fundamental conflict: "nature-technology collisions". This phenomenon reflects the challenges and tensions that arise when technological innovations interfere with natural ecosystems, altering ecological balances and affecting biodiversity. This essay analyzes the consequences and risks of these collisions, exploring the environmental, social and ethical impacts of the relationship between nature and technology.

The Impact of Technology on Natural Ecosystems

Collisions between nature and technology become particularly evident when human activities, driven by technological innovations, alter or destroy natural ecosystems. Deforestation, air and water pollution, overexploitation of natural resources and climate change are just some of the negative side effects of technologies that do not adequately consider environmental constraints. According to the United Nations Environment Programme (UNEP) 2020 report, the expansion of technology, especially industrialization and urbanization, has accelerated the degradation of vital ecosystems such as tropical forests, oceans and wetlands.

For example, large-scale mining, used to extract materials needed for the production of electronic devices and renewable energy, has a devastating impact on biodiversity. Mining of lithium, an essential component in cell phone batteries and electric vehicles, has resulted in the destruction of natural habitats, especially in sensitive regions such as the Atacama Desert in Chile (Liu et al., 2020). The extraction of this mineral has contaminated water sources and affected local flora and fauna, thus altering the ecological balance of the region.

Electronic Pollution and Environmental Degradation

One of the greatest risks arising from the relationship between technology and nature is electronic pollution, also known as "e-waste". The proliferation of obsolete or discarded electronic devices generates enormous amounts of waste, containing toxic substances such as lead, mercury and cadmium. These materials can leach into the soil and water sources, affecting the health of ecosystems and, ultimately, human life.

Electronics recycling is still a limited practice in many parts of the world, and a large amount of e-waste ends up in landfills or in developing countries, where it is informally processed under hazardous conditions. A study by the United Nations University (2017) estimated that only 20% of e-waste is properly recycled, leaving the remaining 80% to pollute the environment. This pollution not only affects local ecosystems, but also contributes to climate change due to the greenhouse gas emissions generated in the production and disposal process of these devices.

The Renewable Energy Paradox: Technologies for a Sustainable Future with Environmental Impacts

The drive towards renewable energies, particularly solar and wind power, is one of the most important responses to climate change and the need to reduce carbon emissions. However, even these technologies can create collisions with nature. Large solar installations require large tracts of land, which can lead to the destruction of natural habitats, especially in desert areas or sensitive ecosystems. Similarly, wind farms, which are rapidly being deployed to generate clean energy, can affect migratory birds and other local species due to collisions with turbines.

In addition, the production of solar panels and wind turbines relies on materials such as silicon, lithium and cobalt, the extraction of which also has significant environmental consequences, such as pollution and ecosystem degradation (Cui et al., 2017). Although renewable energies are a promising solution to reduce the global carbon footprint, the way they are produced and implemented should also carefully consider the ecological impacts of mining and manufacturing these devices.

Digital Technologies and Nature's Performance

Digital technologies, such as the internet of things (IoT), artificial intelligence (AI) and automation, are playing an increasingly important role in environmental management. However, these technologies can also contribute to new forms of natural resource exploitation. Data centers hosting cloud services, for example, consume large amounts of energy and require constant cooling, which in turn increases demands on natural resources such as electricity and water. In addition, the manufacture of technology components, including semiconductors, relies on rare earth resources and minerals, contributing to destructive mining and resource depletion (Bardi, 2014).

On the other hand, digital technologies can also help environmental conservation by providing tools for monitoring ecosystems, optimizing natural resources, and improving energy efficiency. For example, real-time sensors and AI systems can predict weather patterns, control air quality, and monitor the health of forests and oceans. However, the balance between the use of these technologies for environmental good and the negative impacts of their production and operation remains a critical issue.

Reconciling Technology and Nature: Pathways to Sustainability

Collisions between nature and technology are not necessarily inevitable. There is growing recognition of the need to integrate more sustainable approaches into the development and implementation of new technologies. Ecological engineering, which seeks to design technologies that respect the limits of ecosystems, is an emerging discipline that proposes innovative solutions, such as the design of biodegradable electronic devices or the development of more responsible mining techniques. Likewise, the concept of circular economy, which promotes the reuse of resources, recycling and waste reduction, is gaining ground as a key strategy to mitigate the environmental impacts of technology (Geissdoerfer et al., 2017).

In the field of renewable energy, research is focusing on improving the efficiency of energy production, reducing negative impacts on ecosystems and developing more sustainable materials for the manufacture of solar panels and wind turbines. Collaboration between scientists, engineers and environmentalists will be essential to ensure that the technologies of the future are truly sustainable.

Chapter 4: Towards a Sustainable Future

Clean Energy and Green Algorithms

In the context of growing concern about climate change and sustainability, the concept of a sustainable future has become crucially important. In this scenario, clean energies and green algorithms emerge as key elements in the transition to a more environmentally friendly society. Clean energies, such as solar, wind, hydro and biomass, are fundamental to reduce greenhouse gas emissions, while green algorithms are presented as technological tools to optimize the use of energy resources and minimize the environmental impacts of technologies. This essay explores the intersection between clean energy and green algorithms, analyzing their potential, challenges and contributions to the development of a more sustainable future.

Clean Energy: Pillar of the Energy Transition

Clean energies are those obtained from renewable and non-polluting sources, which make them a key alternative to fossil fuels. These sources include solar, wind, hydroelectric and geothermal energy, among others. The main advantage of clean energies is that they do not emit greenhouse gases (GHG) during their production, which helps mitigate climate change. According to a report by the International Energy Agency (2021), the massive adoption of clean energies is essential to achieve the global climate goals established in the Paris Agreement.

One of the most significant advances in the field of clean energy has been the reduction in the cost of solar and wind technologies. The cost of solar PV has decreased by more than 80% since 2010, making this energy source a viable option for a growing number of countries and companies. Similarly, wind energy costs have also experienced a considerable drop, which has enabled the expansion of both onshore and offshore wind farms.

Despite these advances, clean energy presents challenges, particularly in terms of intermittency and storage. Solar and wind power depend on specific weather conditions, which can cause energy production to fluctuate. Energy storage in batteries, such as lithium batteries, is essential to ensure a constant and stable supply, but the production of these batteries also poses environmental challenges, such as resource depletion and e-waste pollution (Harvard Kennedy School, 2020).

Green Algorithms: Optimization and Technological Sustainability

Green algorithms are those designed to improve energy efficiency and reduce the environmental impact of technological systems. These algorithms can be applied in a variety of sectors, from optimizing energy use in industry to improving the efficiency of transportation systems and power grid management. The concept of green algorithms is particularly relevant in a world where energy demand continues to increase, and technologies must adapt to use resources more efficiently.

One of the clearest examples of green algorithms is the optimization of smart grids. These systems use algorithms to manage and distribute energy more efficiently, adjusting supply and demand in real time and using renewable energy sources more effectively. Algorithms are also critical in the transportation sector, where they are used to improve the efficiency of electric vehicles and public transportation systems, thereby reducing the carbon footprint associated with transportation.

In addition, artificial intelligence (AI) and machine learning are increasingly being used to create solutions that optimize energy consumption in buildings, industries and entire cities. For example, AI algorithms can predict energy demand and adjust heating, ventilation and air conditioning (HVAC) systems in real time, which can significantly reduce energy consumption. A study conducted by Stanford University (2020) showed that implementing AI in energy management in commercial buildings can reduce energy consumption by up to 30%.

Synergy between Clean Energy and Green Algorithms

The combination of clean energy and green algorithms has the potential to transform the energy sector and accelerate the transition to a low-carbon economy. Green algorithms can improve the efficiency of renewable energy sources, maximize their production and storage, and reduce energy losses in transmission and distribution. For example, AI-based forecasting algorithms can predict wind and solar energy production, allowing power grids to adjust more efficiently to supply and demand.

In addition, green algorithms can optimize the use of renewable energy in industry and construction. Smart buildings, which use sensors and algorithms to adjust energy consumption, are an example of how technology can reduce the carbon footprint of infrastructure. In agriculture, green algorithms can also be used to optimize the use of resources such as water and energy, thus reducing the environmental impact of food production.

Challenges and Barriers to the Implementation of Green Algorithms

Despite the enormous potential of clean energy and green algorithms, there are several challenges and barriers that must be overcome to achieve widespread adoption. One of the main obstacles is the lack of adequate infrastructure to efficiently integrate renewables into power grids. Although renewables have experienced significant advances in terms of cost and generation capacity, the transmission and storage infrastructure is still insufficient to manage the variability and intermittency of these energy sources.

On the other hand, green algorithms, while promising, also require large amounts of data and powerful computational capabilities. Training AI and machine learning algorithms requires large amounts of energy, which can offset the energy efficiency benefits if not managed correctly. In addition, implementing these algorithms in key sectors, such as transportation and industry, can involve high upfront costs and the need to change existing infrastructures.

Policies and Strategies to Promote a Sustainable Future

To ensure a sustainable future, it is necessary for both governments and companies to adopt policies and strategies that promote the use of clean energy and the implementation of green algorithms. At the governmental level, tax incentives and subsidies for the development of clean technologies must be implemented, as well as regulations that encourage energy efficiency and emissions reduction. International collaboration will also be key to addressing the global challenges of climate change and sustainability.

Companies, for their part, must invest in research and development of technologies that use green algorithms to optimize their processes and reduce their carbon footprint. The adoption of circular economy principles, which promote the reuse and recycling of resources, will also be essential to achieve a positive long-term impact.

Is Responsible AI Possible?

Artificial intelligence (AI) has rapidly transformed diverse sectors, from healthcare and education to transportation and finance. As the capabilities of AI continue to expand, so do concerns about the potential risks and negative effects associated with its use, especially when it is implemented irresponsibly or without considering the ethical, social and legal implications. In this context, a crucial question arises: **is it possible to develop responsible AI?** This essay explores the challenges and opportunities around the creation of ethical AI, considering responsibility in the design, implementation and monitoring of intelligent systems, as well as the social implications of their use.

The Definition of Responsible AI

To understand whether responsible AI is possible, it is essential to define what is meant by "responsibility" in this context. Responsible AI is characterized by being transparent, fair, safe, respectful of human rights, and able to operate within the ethical boundaries set by society. This implies that AI systems must not only operate efficiently, but also in a way that promotes human welfare and respects social, cultural and legal norms.

A fundamental principle of responsible AI is **transparency**. This refers to the ability to understand how algorithms make decisions, allowing users, regulators and other stakeholders to review and audit AI decision processes. **Fairness** is another essential principle, as AI must be designed in a way that minimizes bias and discrimination, ensuring that its decisions do not unfairly favor certain groups or individuals over others.

Finally, **security** is a crucial pillar for responsible AI. Intelligent systems must be resilient in the face of attacks, manipulations or failures that could jeopardize the integrity of the system or the users. In addition, **privacy protection** and **respect for human rights** are fundamental requirements to ensure that AI operates ethically and does not infringe on the rights of individuals.

Challenges for Responsible AI

Despite the clarity of these principles, the creation of responsible AI presents numerous technical, ethical and social challenges.

1.1. Algorithmic Biases

One of the main ethical issues in AI is algorithmic bias. AI algorithms learn from large volumes of data, and if this data contains bias, AI systems can replicate and amplify it. This can result in unfair decisions, such as discrimination in employment, criminal justice, or access to medical services. For example, it has been shown that algorithms used in facial recognition systems can be more inaccurate when identifying people of non-white races, which has led to concerns about racial discrimination (Buolamwini & Gebru, 2018).

The challenge here is to develop mechanisms to detect and correct for these biases. Researchers and AI developers must ensure that the data used to train systems are representative, inclusive, and fair. However, this is not a simple task, as biases can be deeply embedded in social and cultural structures, making their total elimination extremely difficult.

1.2. Lack of Transparency and Explainability

More advanced AI systems, especially those based on deep learning techniques, are often considered "black boxes" due to a lack of clear understanding of how they arrive at their decisions. This lack of transparency can lead to distrust and limit the ability of users to challenge or correct erroneous decisions. **Explainability** of algorithms is a key issue in responsible AI, as users must be able to understand how decisions are made, especially when these decisions affect their lives in a meaningful way.

Recent research has shown that the ability to explain an AI's decisions not only improves public confidence, but also enables error correction and identification of potential biases (Gilpin et al., 2018). However, achieving explainability in highly complex AI systems remains a major technical challenge.

1.3. Safety and Autonomy of AI Systems

As AI systems become more autonomous, concerns arise about how to ensure their safety and avoid potential harm. AI algorithms can learn to adapt and modify their behavior based on new data, raising questions about how to ensure that their decisions are always aligned with human values and do not deviate in unexpected ways.

The concept of **explanatory AI** and **value alignment** refers to the need to ensure that AI systems not only act safely, but also act in accordance with the interests and desires of humans. This becomes even more critical in areas such as autonomous driving, where decisions made by an algorithm can have direct consequences on human life.

In addition, AI systems can be subject to cyber-attacks, which increases the risk of them being manipulated to act in a harmful way. AI cybersecurity is therefore a key aspect of building responsible AI.

Strategies for Developing Responsible AI

Although the challenges to developing responsible AI are significant, there are several strategies that can help overcome these obstacles.

1.1. Development of Ethical Standards and Regulations

One of the most effective ways to ensure responsible AI is through the implementation of ethical rules and regulations. The European Union has led in this regard with the proposal of the **Artificial Intelligence (AI) Act**, which establishes a regulatory framework for the use of AI in high-risk applications, requiring transparency, explainability and accountability in AI systems used in areas such as health, justice and public safety (European Commission, 2021).

These regulations should ensure that AI developers follow clear ethical principles and that systems are auditable and verifiable. In addition, they should encourage the creation of independent **regulatory bodies** to oversee the development and implementation of AI technologies.

1.2. Multidisciplinary Collaboration

The development of responsible AI requires collaboration between experts in AI, ethics, law, sociology and other disciplines. This multidisciplinary approach can help identify and address ethical and social risks before systems are deployed on a large scale. Ethics committees and public consultations are also critical to ensure that AI systems are developed in accordance with societal values and needs.

1.3. Promoting Corporate Social Responsibility

Companies that develop and commercialize AI technologies also have a crucial role to play in promoting responsible AI. Companies must integrate ethical principles into their research and development processes, as well as into business decision-making. This involves not only compliance with existing regulations, but also the adoption of **corporate social responsibility (CSR)** practices that prioritize human and social welfare over immediate profits.

Revolution or Renunciation: Choose an Ethical Path

The intersection between technology, ethics and society is one of the greatest challenges of our era. Constant technological innovation and advances in fields such as artificial intelligence, biotechnology, robotics and renewable energies, for example, offer the promise of progress and prosperity. However, these innovations also bring with them a series of ethical dilemmas and social risks that invite us to ask ourselves: should we embrace these advances with optimism or forgo them because of their possible negative consequences? In this context, a fundamental question arises: **is an ethical and responsible technological revolution possible?** This essay explores the different perspectives on this question and offers an analysis of how we can choose an ethical path in the midst of accelerating technological evolution.

The Promise of the Technological Revolution

The **technological revolution** has been, and continues to be, a key driver for the development of modern societies. In healthcare, for example, technological innovations such as **personalized medicine**, **gene therapies** and **precision treatments** have improved the life expectancy and quality of life of millions of people. Advances in **renewable energy** promise a transition to a more sustainable energy future, and **artificial intelligences** are transforming entire industries, from manufacturing to education to scientific research.

Proponents of the technological revolution argue that innovation is the main driver of human progress. They see technology as having the potential to solve many of humanity's most pressing problems, such as climate change, resource scarcity and incurable diseases. According to philosopher and futurist **Ray Kurzweil**, technology can be seen as a catalyst for reaching the **singularity**, a time when artificial intelligence will surpass human intelligence, offering unprecedented possibilities for the improvement of human life and the advancement of science and understanding of the universe (Kurzweil, 2005).

The Risks of the Technological Revolution: Ethical and Social Dilemmas

However, the rapid adoption of new technologies poses profound ethical risks and dilemmas. Chief among these is the social and economic impact of automation and AI. While technologies can improve productivity and quality of life, they can also lead to **mass unemployment** and growing **economic inequality**, as traditional jobs in sectors such as manufacturing, transportation and agriculture may be replaced by intelligent machines.

In the field of **biotechnology**, advances in gene editing, such as CRISPR technology, offer the possibility of eradicating inherited diseases and improving quality of life. However, these advances also raised fears about the ethical limits of human genetic modification. Modern **"eugenics,"** the idea of improving human genetics to select for desired characteristics, raises questions about the **alteration of humanity** and the risk of creating a **society divided** between those with access to advanced technologies and those without.

Climate change, another global challenge, highlights the duality of the technological revolution. While green technologies and **clean energy** promise to mitigate the effects of global warming, the same technologies can have negative impacts on the environment, such as the extraction of rare minerals needed for electric vehicle batteries or solar panels. In addition, technological solutions often require large amounts of energy, which could exacerbate the problem if not properly managed.

AI algorithms, on the other hand, can be used to make decisions in areas as diverse as criminal justice, labor recruitment, or health care resource allocation. However, these systems can reproduce and amplify existing biases, perpetuating social injustices or historical discriminations, as observed in the case of facial recognition systems that have a higher margin of error in people of non-white races (Buolamwini & Gebru, 2018).

The Waiver Alternative: Caution as an Ethical Pathway

Faced with these dilemmas, some argue that, rather than allowing unbridled adoption of new technologies, we should adopt a more **precautionary** approach. This "forbearance" approach does not mean rejecting innovation altogether, but rather acting prudently and carefully assessing the risks and social implications before adopting new technologies on a large scale.

This approach is advocated by philosophers and activists who believe that technological progress does not always guarantee the well-being of humanity. According to philosopher **Ivan Illich**, the excess of technology can result in **human alienation**, where people lose control over their lives and become mere cogs in a machine that subjugates them (Illich, 1973). The rapid expansion of technology can dehumanize society and generate dependencies that make it more vulnerable to economic, political or environmental crises.

A clear example of this cautious approach is the movement against **uncontrolled artificial intelligence**, which warns of the existential risks posed by out-of-control AI, such as the creation of machines with superior capabilities to humans that could act against the interests of humanity. **The risk of an AI that escapes human controls** is a concern expressed by futurists such as **Nick Bostrom**, who points out that a superintelligent AI could make autonomous decisions without regard for human values, which could have catastrophic consequences (Bostrom, 2014).

The **disclaimer**, then, is based on the need to recognize the limits of what technology can and should do, as well as the need to avoid **technological progress at any cost**, especially when the risks outweigh the benefits.

An Ethical Path: Balancing Revolution and Caution

Rather than choosing between unbridled technological revolution and total renunciation, a viable ethical approach might be to **balance innovation with caution**. This balance involves accepting the potential of technology to improve human life, but also recognizing its inherent risks and acting to minimize them.

This requires the development of **robust ethical and regulatory frameworks** to guide the development and implementation of new technologies. This includes the creation of **clear and stringent regulations** for AI, biotechnology and other emerging technologies, as well as the promotion of an **open and multidisciplinary dialogue** on the social, economic and cultural implications of the technology.

Furthermore, it is essential that decisions on technological adoption be made in a **participatory** and inclusive manner, involving all stakeholders, including governments, businesses, civil organizations and the general public. Only in this way will it be possible to make informed decisions that ensure that technology is used in an ethical manner and that it benefits all members of society, avoiding the dangers of concentrating power and resources in the hands of a few.

Chapter 5: Final Reflection

The Human and Environmental Cost of Progress

Progress, understood as the continuous improvement of human living conditions through science, technology and innovation, has been one of the main drivers of modern societies. Since the Industrial Revolution, humanity has experienced significant advances in various fields: medicine, technology, agriculture and infrastructure. These advances have led to a substantial improvement in the quality of life for billions of people, facilitating increased life expectancy, access to education, global connectivity and economic well-being. However, every step forward has come at a price. In this context, a central question arises: **what is the human and environmental cost of progress?** This essay reflects on the negative effects of development on both humans and the planet, and examines whether it is possible to strike a balance between progress and sustainability.

The Human Cost of Progress

One of the most obvious aspects of technological progress is the **human cost** associated with its development and expansion. Despite improvements in quality of life, technological advances often have a direct impact on people's lives, especially on those who are displaced or disadvantaged by the change.

1.1. Inequality and Labor Displacement

Technological progress has altered traditional labor structures. Automation, robotization and artificial intelligence have made previously unthinkable levels of efficiency possible. However, this automation has brought with it job losses in key sectors, especially in manufacturing and agriculture. According to a report by the International Labor Organization (ILO), more than **100 million workers** worldwide could be displaced due to automation in the coming years (ILO, 2021).

This massive displacement generates growing **economic inequality**, where one part of the population benefits from technological advances, while another is relegated to unemployment or precarious jobs. Automation, for example, has created a **digital divide**, where people without access to technological education or the necessary tools to adapt are excluded from the new labor markets.

In addition, **precarious work** associated with the digital economy, such as work on delivery platforms or autonomous driving, has generated a new form of labor exploitation. Although these activities may seem innovative and modern, they are often characterized by low wages, lack of employment benefits and vulnerability to sudden changes in the market.

1.2. Health and Wellness

Progress has also brought with it new risks to human health. Infectious diseases, such as global pandemics, have increased in frequency, in part due to increased global mobility and digital interconnectedness. Working conditions in high-tech factories, long hours in front of screens and the stress associated with rapid advances have also had detrimental effects on people's mental and physical health.

The increased use of **toxic chemicals** in industrialized agriculture and mass production has also contributed to chronic diseases in populations living near industrial areas. In fact, studies have linked the massive use of pesticides to increased cases of **cancer**, **respiratory diseases** and fertility problems.

The Environmental Cost of Progress

The environmental impact of technological progress is perhaps the most visible and urgent cost. Although progress has enabled a substantial improvement in human life, it has also had a destructive effect on the environment, endangering biodiversity, ecosystems and the global climate.

1.1. Environmental Degradation and Resource Exploitation

The industrial development model has historically relied on the **massive extraction of natural resources**. From mining to deforestation, resource extraction has been driven by the need to fuel industries and satisfy global consumption. **Fossil fuels**, such as oil, gas and coal, have been the main drivers of economic and energy progress, but their burning has led to **unprecedented climate change**, with rising global temperatures and extreme weather events.

Deforestation has been another devastating effect of industrial progress, with millions of hectares of tropical forests destroyed to make way for industrialized agriculture and urbanization. This not only destroys natural habitats, but also contributes to the loss of biodiversity, as many plant and animal species cannot survive the destruction of their natural environment.

In addition, **pollution** from industrial waste, plastics and chemicals has affected aquatic and terrestrial ecosystems, contaminating rivers, seas and soils. It is estimated that around **8 million tons of plastic** are produced each year and end up in the oceans, posing a direct threat to marine life and, ultimately, to humans who depend on these resources for food.

1.2. Impact of Technological Progress in Energy

One of the most celebrated developments in recent decades is the transition to **renewable energies**, such as solar, wind and hydroelectric, which promise to reduce the planet's carbon footprint. However, the production of the technologies needed for these energies, such as solar panels and wind turbines, also comes at an environmental cost. The **extraction of rare minerals** for the manufacture of batteries and electronic components, and the environmental impact of large hydroelectric dams, remain critical issues.

Moreover, the **energy transition** is not without its challenges, as some countries and communities still rely heavily on fossil fuels. This creates a **global mismatch** in the capacities of countries to deal with climate change in an equitable manner, raising questions about climate justice.

1.3. Is Sustainable Progress Possible?

Faced with these costs, the question arises as to whether it is possible to move toward progress that is **sustainable**, both for human beings and for the planet. Progress should not be measured only in terms of economic growth or technological advances, but in terms of **social welfare**, **economic justice** and **environmental balance**.

1.4. Circular Economy and Sustainability

One of the most promising solutions is the adoption of a **circular economic model**, which seeks to reduce waste and promote the recycling and reuse of materials. This model proposes a continuous cycle in which products and materials are kept in use for as long as possible, avoiding the exploitation of non-renewable resources and reducing the amount of waste.

1.5. Clean Technologies and Renewable Energy

The future of progress can be driven by **clean technologies** and **renewable energy**. While the transition to a green economy presents challenges, it also offers the opportunity to create sustainable jobs, improve air quality and reduce the effects of climate change. Investing in research and development of technologies that are not only efficient but also environmentally friendly will be crucial to ensure a more balanced future.

At What Price the Future?

The question of the price of the future is one of the most complex and pressing questions facing modern societies in the context of accelerating technological progress, socioeconomic change and global environmental challenges. As humanity moves towards a seemingly promising future, driven by innovation, new technologies and economic growth, fundamental questions arise about the costs that this progress entails: Are we willing to pay the price of a more technological future at the cost of our social relations, economic equity, environmental justice and, ultimately, our humanity? This essay will reflect on the costs inherent in the future we are building and how they affect different aspects of human and planetary life.

The Promised Future: Progress and Expectations

In recent decades, technological advances have given rise to a future that, in many respects, seems ideal. Emerging technologies, from artificial intelligence (AI) to biotechnology and nanotechnology, promise to transform medicine, education, transportation and communications. Sustainable energy solutions, such as renewable energy, offer the possibility of a future without dependence on fossil fuels, while innovations in agriculture , such as **precision farming** and **genetically modified crops**, could feed an expanding global population without destroying the environment.

On a societal level, the digital future also brings with it **global connectivity** and near-instant **access to information**. The integration of emerging technologies into everyday life is reshaping education, enabling more accessible, personalized and globalized learning, while automation promises to free humans from repetitive and dangerous tasks.

At the same time, however, these advances are taking place in a context of social inequality, exploitation of natural resources and pressure on the environment. What are we willing to sacrifice in the name of progress?

The Social Cost of the Technological Future

1.1.Social Inequality and Digital Divide

One of the most obvious costs of the technological future is increased **social inequality**. While emerging technologies have the potential to improve the quality of life for millions, the unequal distribution of these technologies can accentuate existing disparities. In the digital world, the

digital divide between those who have access to high-speed internet, advanced technological devices and digital education, and those who do not, is a key factor in perpetuating inequality.

In addition, automation and AI may lead to **mass unemployment**, especially in sectors that rely on manual labor or routine jobs. The **displacement of workers** by artificial intelligence, especially in countries with low-wage economies, may lead to a future of **job instability** and **social unprotection** for millions of people.

In this sense, the central issue is not only the creation of new technologies, but also **how to distribute their benefits**. If technological progress is not managed equitably, the future could be marked by **increased poverty** and **social polarization**, where only a small technological elite enjoys the benefits of progress, while the rest of the population is left behind.

The Impact on Human Relationships

In addition to the economic aspects, technological progress is changing the nature of **human relationships**. Social networks, instant messaging applications and digital platforms have transformed social interactions, but not without costs. Interpersonal relationships have become more superficial and fragmented, and the increasing reliance on technology for communication can lead to **social isolation**.

Technology platforms have also created a world in which **privacy** has been undermined, where personal data is constantly collected and sold without the explicit consent of users. This control of data and vulnerability of **personal information** becomes a price that people must pay to access the benefits of technology.

The Environmental Cost of the Future

3.1. Exploitation of Natural Resources

The future we are building is deeply tied to the exploitation of natural resources. Despite advances in **renewable energy**, the global demand for **rare minerals** needed to manufacture technological devices such as cell phones, electric batteries and solar panels is putting pressure on ecosystems and increasing mining in ecologically sensitive locations. The **unbridled exploitation of** natural **resources** raises the question of whether we can sustain long-term technological growth without depleting the planet's resources.

The energy transition to cleaner sources is necessary, but it is not without its own ecological costs. The production of lithium batteries, for example, requires large amounts of water and can cause significant **environmental damage** if not properly managed. Thus, the development of new "green" technologies is not always as environmentally friendly as it first appears.

3.2. Pollution and Technological Waste

Planned obsolescence and mass production of electronic devices generate an **accumulation of electronic waste** that affects ecosystems. These devices contain toxic substances such as **mercury**, **cadmium** and **lead**, which can leach into the soil and water, harming fauna and flora. The management of this waste has become a global challenge, with developing countries often becoming the dumping grounds for these obsolete products.

Although solutions to this problem, such as the recycling of electronic components, are underway, large-scale solutions are costly and still insufficient to mitigate the impact of the **"digital footprint"**.

The Future of the Future: At What Price?

The final reflection on the price of the future should lead us to consider not only technological advances and solutions, but also the **ethical, social and environmental costs** they entail. Technological progress is not a neutral process: there are always decisions behind each innovation, decisions that affect people, nature and the balance between the two.

The key is that, as a society, we must be aware of the **prices** we are willing to pay. This implies:

- Develop **sound ethical frameworks** to guide the development and implementation of new technologies.
- Ensure that **technologies are accessible and equitable** for all, avoiding the widening of existing inequalities.
- Prioritize **environmental sustainability** in the design and production of technologies, ensuring that natural resources are used responsibly.
- To foster a **culture of social and ethical responsibility** in technological innovation, where the well-being of people and the planet is the main objective.

If we do not stop to reflect on **at what cost we are building the future,** the future we imagine could be much less promising than we expected.

Epilogue:

As we move toward a future increasingly dominated by artificial intelligence, the fundamental question persists: At what price are we willing to accept progress? We have explored the glowing horizon that AI promises, but we have also touched on the shadows cast behind its advance. From the invisible pollution of data centers to the ethical imbalances that arise from algorithms designed without conscience, the cost of our technological dependence is indisputable.

However, the challenge lies not only in the problems we have identified, but in our ability to address them in an ethical and sustainable manner. AI has transformative potential that could improve human life, but only if we choose a path that does not ignore the impact on the environment, human rights or social justice. The future of artificial intelligence is not set in stone, and it is up to us to decide how to balance its innovations with a holistic view of what progress really means.

The responsibility does not lie solely with developers and large technology corporations; each of us, as users, consumers and citizens, also has a crucial role to play. The key will be in our ability to advocate for a conscious and ethical use of AI, pushing for policies that promote transparency, sustainability and fairness.

The challenge is great, but not insurmountable. The future, though uncertain, is full of opportunities to reimagine a world where technology and nature are not seen as enemies, but as allies in building a more just and sustainable society. The artificial intelligence revolution is inevitable, but how we choose to navigate it will determine whether we are truly capable of achieving the future we desire: one in which technology serves human and environmental well-being equally.

Bibliographic references

Akenji, L. (2014). *The Sustainability of Obsolescence: Impacts of Short Life-Cycles on Consumers and the Environment*. Springer.

International Energy Agency (2022). *Data Centres and Data Transmission Networks*. Retrieved from https://www.iea.org

Andrae, A. S. G., & Edler, T. (2015). On Global Electricity Usage of Communication Technology: Trends to 2030. *Challenges, 6*(1), 117-157. https://doi.org/10.3390/challe6010117

Amnesty International (2016). *This is what we die for: Human rights abuses in the Democratic Republic of the Congo power the global trade in cobalt*. Retrieved from https://www.amnesty.org

Angwin, J., Larson, J., Mattu, S., & Kirchner, L. (2016). Machine Bias: There's Software Used Across the Country to Predict Future Criminals. And It's Biased Against Blacks. *ProPublica*.

Apple (2022). *Annual Financial Report*. Retrieved from https://www.apple.com

Babbitt, C. W. (2017). Environmental Impacts of Electronic Waste: A Global Overview. *Environmental Science & Technology, 51*(4), 2381-2390.

Inter-American Development Bank (2024). *Artificial intelligence: what it brings and what it changes in the world of work*. Retrieved from blogs.iadb.org

World Bank (2020). *Minerals for Climate Action: The Mineral Intensity of the Clean Energy Transition*. Retrieved from https://www.worldbank.org

Baldé, C. P., Forti, V., Gray, V., & Kuehr, R. (2017). *The Global E-waste Monitor 2017: Quantities, Flows, and the Circular Economy Potential*. United Nations University.

Bostrom, N. (2014). *Superintelligence: Paths, Dangers, Strategies*. Oxford University Press.

Bostrom, N. (2023). "Artificial intelligence will be the last invention mankind will ever make". In *Rísbel Magazine*. Retrieved from https://risbelmagazine.es

Buolamwini, J., & Gebru, T. (2018). Gender Shades: Intersectional Accuracy Disparities in Commercial Gender Classification. *Proceedings of the 1st Conference on Fairness, Accountability, and Transparency*.

BBC News. (2021). *Facebook data on 530 million users found online*. Retrieved from https://www.bbc.com

Cadenas, B. (2024). *Impact of artificial intelligence in science, industry and society. Gaceta UNAM*. Retrieved from gaceta.unam.mx

Chan, J., Pun, N., & Selden, M. (2013). The Politics of Global Production: Apple, Foxconn, and China's New Working Class. *Asia-Pacific Journal: Japan Focus, 11*(45), 1-41.

European Commission (2021). *Proposed Artificial Intelligence Act: A framework for ensuring trusted AI in Europe*. European Commission.

Cohen, M. J. (2017). Understanding Consumer Demand in a Circular Economy: The Role of Obsolescence in the Technology Market. *Journal of Industrial Ecology, 21*(4), 1012-1021.

Crawford, K. (2023). In *Entrepreneurs News*. Retrieved from https://emprendedoresnews.com

Cui, L., Zhang, D., & Zhao, H. (2017). Environmental Impacts of Lithium Extraction for Electric Vehicles and Solar Technologies. *Environmental Science & Technology, 51*(23), 14101-14109.

DeepMind (2018). DeepMind AI Reduces Google Data Centre Cooling Bill by 40%. Retrieved from https://deepmind.com

Dastin, J. (2018). Amazon Scraps Secret AI Recruiting Tool That Showed Bias Against Women. *Reuters*.

EITB (2024). *The environmental impact of the use of artificial intelligence*. Retrieved from eitb.eus

European Commission (2018). *Directive 2012/19/EU on Waste Electrical and Electronic Equipment (WEEE)*. Official Journal of the European Union.

Eubanks, V. (2018). *Automating Inequality: How High-Tech Tools Profile, Police, and Punish the Poor*. St. Martin's Press.

Geissdoerfer, M., Savaget, P., Bocken, N. M. P., & Hultink, E. J. (2017). The Circular Economy - A New Sustainability Paradigm?. *Journal of Cleaner Production, 143*, 757-768.

Gilpin, L. H., Bau, D., Yuan, B. Z., & Vishnu, A. (2018). Explaining Explanations: An Overview of Interpretability of Machine Learning. *ACM Computing Surveys, 51*(5), 1-42.

Google (2020). *Our Commitment to Sustainability*. Retrieved from https://sustainability.google

Harvard Kennedy School (2020). *The Future of Energy: Transitioning to Renewable Sources*. Harvard University Press.

Harvard University (2020). *The Ethics of Artificial Intelligence: Implications for Society*. Harvard Kennedy School.

Illich, I. (1973). *Deschooling Society*. Harper & Row.

International Energy Agency (IEA). (2021). *World Energy Outlook 2021: Accelerating Clean Energy Transitions*. IEA.

Kurzweil, R. (2005). *The Singularity Is Near: When Humans Transcend Biology*. Viking.

La Vanguardia (2023). *AI: Energy consumption and carbon footprint*. Retrieved from lavanguardia.com

Leung, A. O. W., Zhou, B., & Wong, M. H. (2015). Electronic Waste Recycling: A Critical Review. *Environmental Science & Technology, 49*(11), 6824-6841.

Li, F.-F. (2023). "AI should be seen as a tool that extends our cognitive capabilities, not as a substitute for human intelligence". In *Rísbel Magazine*. Retrieved from https://risbelmagazine.es

Liu, Z., Smith, S., & Shuman, J. (2020). The Environmental Impact of Lithium Mining in South America: A Review of Sustainability. *Environmental Research Letters, 15*(6), 063001.

Musk, E. (2023). "If AI has a goal and humanity happens to be in the way, it will destroy humanity as a matter of course". In *Rísbel Magazine*. Retrieved from https://risbelmagazine.es

Newton, C. (2019). The Trauma Floor: The Secret Lives of Facebook Moderators in America. *The Verge*. Retrieved from https://www.theverge.com

Obermeyer, Z., Powers, B. W., Vogeli, C., & Mullainathan, S. (2019). Dissecting Racial Bias in an Algorithm Used to Manage the Health of Populations. *Science, 366*(6464), 447-453.

United Nations (2020). *The Global E-waste Monitor 2020*. Retrieved from https://www.ewastemonitor.info

Intergovernmental Panel on Climate Change (2022). *Climate Change 2022: Mitigation of Climate Change*. Retrieved from https://www.ipcc.ch

Perfect Energy (2024). *Mining and resources in the age of artificial intelligence*. Retrieved from perfectaenergia.com

ProPublica (2016). *Machine Bias*. Retrieved from https://www.propublica.org

Schiebinger, L. (2020). *AI and Gender*. Stanford University.

Schwab, K. (2016). *The Fourth Industrial Revolution*. Crown Business.

Slade, G. (2006). *Made to Break: Technology and Obsolescence in America*. Harvard University Press.

Stanford University (2020). *AI for Sustainability: How Artificial Intelligence Can Reduce Energy Use in Buildings*. Stanford News.

Statista (2023). *Global market share of cloud infrastructure services by vendor*. Retrieved from https://www.statista.com.

Stanford University (2020). *Machine Learning and Energy Efficiency: Opportunities for Commercial Buildings*. *Energy Science & Engineering, 8*(2), 189-204.

Strubell, E., Ganesh, A., & McCallum, A. (2019). *Energy and Policy Considerations for Deep Learning in NLP*. Retrieved from arxiv.org.

TechCrunch (2019). *Marketing company exposed millions of personal records*. Retrieved from

The Shift Project (2019). *Climate Crisis: The Unsustainable Use of Online Video*. Retrieved from

Unite.AI (2024). *The impact of AI on innovation*. Retrieved from unite.

University of the United Nations (2017). *The Global E-Waste Crisis: Opportunities and Solutions*. UNU-IAS.

UNEP (2022). *Global E-waste Monitor 2022: Quantities, Flows and the Circular Economy Potential*. United Nations Environment Programme.

World Economic Forum (2023). *Global Risks Report 2023*. WEF.

Zuboff, S. (2019). *The Age of Surveillance Capitalism: The Fight for a Human Future at the New Frontier of Power*. PublicAffairs.

Printed by Books on Demand GmbH, Norderstedt / Germany